Images of War

HITLER'S LIGHT PANZERS

1935 – 1943

Paul Thomas

Pen & Sword
MILITARY

First published in Great Britain in 2015 by
PEN & SWORD MILITARY
an imprint of
Pen & Sword Books Ltd,
47 Church Street,
Barnsley,
South Yorkshire,
S70 2AS

A CIP record for this book is available from the British Library.

ISBN 978 1 78346 325 1

Printed and bound in Malta by Gutenberg Press Ltd

Pen & Sword Books Ltd incorporates the Imprints of
Pen & Sword Aviation, Pen & Sword Maritime,
Pen & Sword Military, Wharncliffe Local History, Pen & Sword Select,
Pen & Sword Military Classics and Leo Cooper.

For a complete list of Pen & Sword titles please contact
Pen & Sword Books Limited
47 Church Street, Barnsley, South Yorkshire, S70 2AS, England

E-mail: enquiries@pen-and-sword.co.uk
Website: www.pen-and-sword.co.uk

Contents

Introduction

Hitler's Light Panzers at War is an illustrated record of the German light tank from its beginnings in the 1930s, to the key battles it fought in Poland, France, North Africa, Russia and north western Europe. The book analyses the development of the light Panzer, which ranged from the Panzers I and II to the Czech-built Panzers 35 and 38(t). It describes how the Germans carefully utilized the development of these light machines for war, and depicts how these tanks were adapted and up-gunned to face the ever-increasing enemy threat.

Using over 200 rare and unpublished photographs together with detailed captions and accompanying text, *Hitler's Light Panzers at War* provides a unique insight into the many variants that saw action on the battlefield. It provides a vivid account of Panzer development and deployment from the early Blitzkrieg campaigns, to the final demise of the Nazi war machine.

Chapter I
Development and Training

During the early 1930s the German Army, which was still limited following the Treaty of Versailles, instructed a number of German firms to fund and design a light and medium tank that would be versatile, strong and reliable on the battlefield. It was also proposed that the light tank would have to be available in large numbers and be financially viable, to be produced quickly and afford good all-round fire power, both in an offensive and a defensive role.

It was agreed that a new light tank was to be designed under a 5-ton weight limit that was capable of serving the new *Panzerwaffe* with a small good all-round tank that could also be used to train Panzer crews. Five German firms submitted their prototype proposals, and from this Krupp were selected to finally produce a light tank. Within months plans were drawn up and funds were made available for a light and a medium tank.

In 1933, Krupp finally delivered their first prototype light panzer – known as the the La.S or *Landwirtschaftlicher Schlepper* – agricultural towing vehicle. The Versailles Treaty was the reason for this misleading name. This new tank had a Daimler-Benz superstructure and turret. The testing of the vehicle quickly proceeded and throughout 1934 it was put through a number of stringent tests in the training grounds. The German Army assigned the designation of the tank as a 'Krupp-Tractor'. By April 1936 it was officially designated Pz.Kpfw.I Ausf.A (Sd.Kfz. 101).

The *Führer*, Adolf Hitler, came to the training grounds to see for himself the new Pz.Kpfw.I, or Panzer I, and immediately told his staff that he envisaged a fast-moving army of tanks that would spread fire and devastation such as the world had never before seen. He made it known that the tank would be the prime machine that would use battlefield tactics, moving with rapid speed, to achieve its objectives quickly and effectively.

The Panzer I Ausf.A variant featured a crew of two, a driver and a commander, the latter also used as the gunner. The driver sat in the forward hull of the cramped vehicle on the left, whilst the commander occupied the turret to the right. The tank was armed with 2 x 7.92mm machine guns, both of which were capable of firing 650 rounds per minute, could be fired simultaneously or individually, and could only be traversed by the commander by hand.

Entry and exit for the commander was through the small turret roof, whilst the driver could exit or enter the vehicle by a hinged rectangular door alongside the left of the superstructure.

The tank had minimal armoured protection and featured five road wheels to a track side and each wheel was encased in rubber. Three rollers were fitted to the underside of the upper track run. Operating weight was listed at 5.9 tons and power came from a single Krupp M 305 air-cooled, four-cylinder petrol engine delivering up to 60 horsepower. The Ausf.A could manage a top on-road speed of 23mph, with an operational range of 85 miles cross country, or 125 miles on road.

The new prototype was regarded as a success, in spite of the fact that the tank had limited battlefield capability. In order to speed the process of manufacture of the new Panzer I Ausf.A other German firms, such as Henschel, Daimler-Benz and MAN, were brought in to support Krupp's production. Henschel received a first batch order to produce 150.

While manufacture of the Ausf.A continued in earnest, in August 1935 a more powerful engined and slightly longer Panzer I was developed. The Ausf.B variant still had a crew of two, but this time it was powered by a single Maybach NL 38 TR six-cylinder, liquid-cooled gasoline engine developing 100 horsepower. By the end of 1937, 399 were delivered, but production of this model ceased in June of that year.

The Panzer I was a promising introduction to what the Germans could achieve in German tank design. In 1936 the tank was used in large scale manoeuvres comprising infantry and Luftwaffe formations. It was in this year, too, that Hitler committed a 'volunteer' army of troops, aircraft and Panzers to aid the Spanish Civil War (1936 – 1939). This was the ideal proving ground for future operations utilizing troop concentration, armoured vehicles and Luftwaffe support.

Another light tank that was prominent in its development in the 1930s was the Czechoslovak tank manufacturer ČKD, which had been looking for a replacement for the LT-35 tank they were jointly producing with their Škoda Works.

On 1 July 1938, Czechoslovakia ordered 150 of the TNHPS model. However, by the time of the German occupation, none had entered service. After the German takeover, the Germans ordered the continued production of the model, as it was considered an excellent tank, especially compared to the Panzer I and Panzer II tanks that were the Panzerwaffe's main tanks. The special vehicle designation for the tank in Germany was the Sd. Kfz. 140. However, it quickly reverted to the Pz.Kpfw 38(t). This riveted armoured, rear-engine tank had a two-man turret, which was centrally located, and housed the tank's main armament, a 3.7mm Skoda A7 gun with 90 rounds stored on board. It was equipped with a 7.92mm machine gun to the right of the main ordnance. This turret machine gun was in a separate ball mount rather than a fixed coaxial mount. The driver was situated in the front right of the hull, with the bow machine-gunner seated to the left, manning the 7.92mm machine gun. The bow gunner also doubled as the radio operator. The radio was mounted on the left of the bow gunner.

The engine was mounted in the rear of the hull and drove the tank through a transmission with five forward gears and one reverse gear to forward drive sprockets. The track ran under four rubber-tyred road wheels and back over a rear idler and two track return rollers. The wheels were mounted on a leaf-spring double-bogie mounted on two axles.

Another Czech design, which was also used as a German light tank following the occupation of Czechoslovakia, was the Lehký tank vzor 35 (Light Tank Model 35) and was designated by the Germans as the Pz.Kpfw 35(t). Four hundred and thirty-four were built. Of these, the Germans seized 244 when they occupied Bohemia-Moravia in March 1939.

The tank had a four-man crew: the driver sat on the right side of the tank using an observation port. The radio operator sat on the left with his own observation port. His radios were mounted on the left wall of the hull. The hull machine gun was between the driver and the radio operator in a ball mount. Most of the machine gun's barrel protruded from the mount and was protected by an armoured trough. In the turret sat the commander, who was responsible for loading, aiming and firing the main gun and the turret machine gun, whilst at the same time commanding the tank.

Another revolutionary light tank built for the German war machine was the *Panzerkampfwagen* II or abbreviated as the Pz.Kpfw.II. The design of the tank was based on the Pz.Kpfw.I, but was larger and with a turret mounting a 20mm anti-tank gun. Production began in 1935, but it was not combat-ready until 1936.

All production variants of the Pz.Kpfw.II were fitted with the 140 PS, gasoline-fuelled six-cylinder Maybach HL 62 TRM engine and ZF transmissions. The Ausf.A,

B and C variants had a top speed of 25mph, while the Ausf.D and E had a torsion bar suspension and a much superior transmission, giving a top road speed of 33mph. However, across country where this vehicle would be used mainly, it had a much lower speed than previous models. Consequently in the Ausf.F variant, the old leaf-spring type suspension was used, making it much faster.

The Pz.Kpfw.II had a three-man crew. The driver sat in the forward hull, whilst the commander sat in a turret seat and was also the gunner. The radio operator was positioned on the floor of the tank under the turret.

Five photographs showing recruits putting the Panzerkampfwagen I A, also known as the LaS, through training exercises in 1934. The crew are demonstrating the versatility of the vehicle by using rough terrain. The open compartment with no turret gave the driver all-round visual and a better understanding of the tanks mobility. Although during training there were sometimes three or even four recruits onboard the vehicle at any one time, the LaS was actually designed for a two-man crew, one serving as the driver and the other as the commander, radioman and machine gunner.

At a Nazi event, the Panzer I, also known as the Pz.Kpfw.I, is revealed to adoring onlookers in 1937 inside a stadium.

A Pz.Kpfw.I during a ceremonial event. The national cross is draped on the front of the vehicle, and its crew, wearing the distinctive black Panzer uniform and Panzer beret, stand either side.

A Panzer unit during a pause in a drill. The crewmen of these Pz.Kpfw.I Ausf.A are wearing the standard black Panzer uniform, a jacket with rose-coloured shoulder boards, collar patches and jacket trim. The black berets were worn over a hard rubber liner to protect the head from injury in the Panzer.

On an exercise are Pz.Kpfw.I with infantry supporting the advance across a field somewhere in Germany in 1938. The Ausf.A variant was nicknamed 'Krupp-Sport' by the soldiers, as it was very noisy on account of its opposed air-cooled engine. The next series, the Ausf. B, however, was slightly quieter, with a water-cooled engine.

At a workshop, engineers pose for the camera with two Pz.Kpfw.I Ausf. B. These vehicles are being prepared to march. The first test for these vehicles was the Spanish Civil War, where they were used by German volunteers of the 'Legion Condor'. The second test – though not with live ammunition – was the march through into the Sudetenland in late 1938.

The crew of a Pz.Kpfw.I Ausf.A pose for the camera during a brief break during a training exercise. The main armament has been removed and placed in storage whilst the vehicle is put through its paces. The man in the turret is the trainer; the other two men are the crewmembers, dressed in the familiar denim training tunics.

Two photographs showing a number of Pz.kpfw.I at a workshop. All of the vehicles are painted in dark grey camouflage and each has been numbered in white stencil on the side of the vehicle.

A line of Pz.Kpfw.I Ausf.A in 1938. The twin MG13 machine guns were removed from all combat vehicles before they were parked. It was also standard practice that all radio equipment was removed from the armoured vehicles.

The crew of a Pz.Kpfw.I Ausf.B show the versatility of the tank by spanning ditches and various other obstacles. This type of training was important, as the tank would be employed on the battlefield in highly mobile operations crossing many different types of varied terrain in order to exploit breakthroughs on the battlefield.

A command Pz.Kpfw.I Ausf.B out in the field, probably during a training exercise. Exercises with the new tanks were undertaken extensively by the *Panzerwaffe* before the war and special attention was made to train forces in close cooperation between the air forces and the Panzers on the ground. It was soon realized that the Panzer had three main tasks: supporting infantry, operating units with other mobile weapons and, finally, combating tanks.

Out in the training ground a Panzer crew are seen posing for the camera with the LaS vehicle. This was a driving school for Panzer recruits.

Here a Pz.Kpfw.I is photographed moving across ground during a training exercise. This vehicle could cross flat dry terrain relatively easily and could be steered without too much trouble. The driver controlled the direction by means of steering levers, each of which had two handgrips, one for normal steering and the other with a thumb-plunger to act as a parking brake.

Female personnel have clambered onto a Pz.Kpfw.I for the camera. This vehicle is more than likely earmarked for Poland in the days or weeks ahead.

A crewmember of a Pz.Kpfw.I Ausf.A poses in front of his vehicle prior to an exercise in early 1939. At the training garrison camps, there were large firing ranges and vast open areas of countryside where these vehicles could be put through their paces.

A Pz.Kpfw.I with a crew member during a training exercise. Here in this photograph the 7.92mm machine gun can clearly be seen. Although it was used extensively for training exercises before the war, it was clearly under-gunned and under-armoured and would not be able to fight effectively against heavier and superior enemy tanks.

Workshop engineers pose for the camera with a Pz.Kpfw.I Ausf.A. These vehicles are being prepared to march.

A posed photograph showing a father and his young child dressed in a Panzer uniform inside the turret of a Pz.Kpfw.I. Prior to war not all tanks had been equipped with radio equipment, and most of the communication in combat training was done with variously coloured signal flags.

At a workshop, engineers are seen making adjustments to one of the training Panzers, the LaS, which was the forerunner of the Pz.Kpfw.I.

During the occupation of the Sudetenland in October 1938, a group of tank men from an unknown tank regiment converse among themselves around a stationary Pz.Kpfw.I Ausf.A. Not a single shot was fired in anger when the Wehrmacht rolled into the Sudetenland. Note the vehicle's MG13 machine guns, still protected by canvas covering the barrels.

Spanning ditches and other obstacles in a Pz.Kpfw.I Ausf.B.

Engineers are seen with a Pz.Kpfw.I out in the field changing one of the vehicle's wheels. This tank is probably being used for training purposes.

A group of Pz.Kpfw.II out in the field. This tank's main armament was a 2cm cannon with a co-axle 7.92mm MG34 machine gun. The 2cm gun was fired from a trigger on the elevating hand wheel to the commander's left, and the MG34 from a trigger on the transverse hand wheel to his right.

A number of Pz.Kpfw.I parading along a road. The tracks of this vehicle had sufficient off-road capability, the ground pressure was extremely good, but because the engine was underpowered it was prone to mechanical problems.

At a training ground is a **Pz.Kpfw.I.** The vehicle consists of overall dark grey with dark green patches over the whole vehicle. **By** 1940 these tanks were mainly finished in overall grey.

On parade is a **Pz.Kpfw.II.** In 1937, the **Ausf.C** became the main prototype with a brand new leaf-spring suspension system with its familiar running gear of five large disc-type road wheels. This variant would later be used not only for combat, but for training purposes as well.

Out in the field during a training exercise. Because there was no radio communication on board these vehicles, a series of signals were made to communicate between the tanks and the trainer.

Moving along a road during training is a column of Pz.Kpfw.I. For identification during training many of these vehicles were nominated single unit numbers, stencilled in white, often on the frontal armour and the sides.

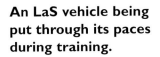

An LaS vehicle being put through its paces during training.

A Panzer commander in the turret of his Pz.Kpfw.I Ausf.B poses for the camera with a smile. He wears the distinctive black Panzer beret that was universally worn in the Panzerwaffe until early 1940. It was replaced by the black M1938 field cap.

A group of Pz.Kpfw.I at a training camp in 1938. By the time war broke out in Poland the following year these vehicles would have their distinctive white painted crosses on the side of the turret. These Panzers were fortunate to be faced with little opposition in Poland.

An LaS vehicle, with a full complement of crew, drives through a local German town during training.

Trainers and their recruits prior to a gruelling exercise on the Lüneburg Heath. In order to prepare the men for battle stations the trainer would blow his whistle to signal the Panzer men to board their vehicles. Getting quickly inside the light tank was practiced frequently by the crews.

A Panzer crew during a pause in a drill. The crewmen of these Pz.Kpfw.I Ausf.A are wearing the standard black Panzer uniform with black Panzer beret.

A Panzer commander is seen sitting in the turret of his Pz.Kpfw.I moving across country.

Two Pz.Kpfw.I during a training exercise, showing the versatility of this vehicle across country. Note the camouflage of the vehicle, which consists of overall dark grey with dark green patches over the whole tank.

A good photograph of a Pz.Kpfw.I during a training exercise. It would not be until the Polish Campaign in 1939 that these Panzer crews realized that these light tanks could not withstand enemy anti-tank fire.

Three photographs taken in sequence showing a crewmember posing for the camera with a **Pz.Kpfw.I** during a training exercise. Inside these vehicles the crew carried some 3,125 rounds of ammunition, which was quite impressive at that time.

At a workshop the crew of a Pz.Kpfw.II pose for the camera. Crews trained extensively with the Ausf.C variant, which soon became known as the 'fast combat wagon'. But still designers were determined to re-develop the vehicle.

An interesting photograph showing a Pz.Kpfw.I towing an LaS trainer through the snow. The LaS has obviously developed a mechanical problem and is being taken back to the workshop for repair.

Chapter II
Blitzkrieg

To carry out the German attack against Poland the Germans had planned for two Army Groups – Army Group North, consisting of the Fourth and Third Armies, under the command of General Fedor von Bock, and the Southern Army Group, consisting of the Eighth, Tenth and Fourteenth Armies, commanded by General Gerd von Rundstedt.

A large part of this huge army was the Panzer, which was to spearhead the attack into Poland. Army Group South had the strongest armoured formations, with over 2,000 tanks and 800 armoured vehicles.

On 1st September 1939, from north to south all five German Army Groups crashed over the frontier. The light Panzers pushed forward with speed, leaving devastation in their wake.

Over the next few days, both the North and South groups continued to make furious thrusts on all fronts. The campaign had taken on the character that was to remain for the few weeks that followed. Everywhere north, south and east, the fronts were shrinking, cracking slowly but surely under the massive German pressure. In this unparalleled armoured dash, some units had covered 40 miles – 60 road miles in just twenty-four hours. For many soldiers it was an exhilarating experience, Panzers bucketing across the countryside, meeting, in some places, isolated pockets of resistance, and destroying them. Despite the determination of the brave Polish soldiers, fast and devastatingly efficient Blitzkrieg had arrived.

The withdrawing Polish Army were being mauled almost to death by constant air attacks and pounded mercilessly by tanks and artillery. The Poles were faced with the finest fighting army that the world had ever seen.

The quality of the German weapons – above all the Panzers – was of immense importance in Poland. Their tactics were the best: stubborn defence; concentrated local firepower from machine-guns and mortars; rapid counter attacks to recover lost ground. Essentially, everything in the invasion went according to plan, or even better than the plan, in the unfolding both of strategy and tactics. Hitler and his generals were overjoyed at the lightening speed and extent of their gains. By the end of September the war in Poland had been won.

Hitler had been so impressed by Blitzkrieg that the following year, in May 1940, he adopted the same tactics against the Low Countries and France. He was resolute that if he was going to win the war rapidly in the west the new Blitzkrieg tactic would be instigated quickly and effectively. Whilst he had been aware that his forces had overwhelming superiority in modern equipment against a country like Poland, he knew that France and her allies had a slight advantage in terms of both numbers of troops and material. Yet, in spite of pessimism from many of his Western Front commanders, he was sure that by adopting the Blitzkrieg tactics of highly mobile operations involving the deployment of motorized infantry, air power, and armour in coordinated attacks, his forces would gain rapid penetration followed by encirclement of a bewildered and overwhelmed enemy.

For the attack against the west the German Army were divided into three army groups – Army Groups A, B and C. The main strike would be given to Army Group A, which would drive its armoured units through the Ardennes, then swing round across the plains of northern France and make straight for the Channel coast, thereby cutting the Allied force in half and breaking the main enemy concentration in Belgium between Army Group A advancing from the south and Army Group B in the north. The task of Army Group B was to occupy Holland with motorized forces and to prevent the linking up of the Dutch army with Anglo-Belgian force. It was to destroy the Belgian frontier defences by a rapid and powerful attack and throw the enemy back over the line between Antwerp and Namur. The fortress of Antwerp was to be surrounded from the north and east and the fortress of Liege from the north-east and north of the Meuse.

Distributed between the three army groups was the armour, which would lead the drive through Belgium, Holland and then into France. In total there would be a staggering 2,072 tanks: 640 Pz.Kpfw.I, 825 Pz.Kpfw.II, 456 Pz.Kpfw.III, 366 Pz.Kpfw.IV, 151 Pz.Kpfw.35(t) and 264 Pz.Kpfw.38(t). The reserves comprised some 160 vehicles to replace combat losses and 135 Pz.Kpfw.I and Pz.Kpfw.II, which had been converted into armoured command tanks, which resulted in them losing their armament. The vehicles that had been distributed among the ten Panzer divisions were not distributed according to formation of the battles they were supposed to perform. The 1 Panzer Division, 2 Panzer Division and 10 Panzer Division each comprised thirty Pz.Kpfw.I, one hundred Pz.Kpfw.II, ninety Pz.Kpfw.III and fifty-six Pz.Kpfw.IV. The 6 Panzer Division, 7 Panzer Division and 8 Panzer Division consisted of ten Pz.Kpfw.I, 132 Pz.Kpfw.35(t) or Pz.Kpfw.38(t) and thirty-six Pz.Kpfw.IV. A further nineteen Pz.Kpfw.35(t) were added to the 6 Panzer Division due to the compliment of a battery of sIG mechanized infantry guns. The 3 Panzer Division, 4 Panzer Division and 5 Panzer Division each consisted of 140 Pz.Kpfw.I, 110 Pz.Kpfw.II, fifty Pz.Kpfw.III and twenty-four Pz.Kpfw.IV.

Yet again, as in Poland, the battle of the Low Countries and then France ended with another victory for the Germans, by June 1940. They had reaped the fruits of another dramatic Blitzkrieg campaign. France had proven ideal tank country to undertake a lightening war, and its conception seemed flawless. It seemed to many of the tacticians that Blitzkrieg would ensure future victories. The Panzer was the key to this success.

A Pz.Kpfw.I tank is being prepared for transportation during manoeuvres in Poland in August 1939. On a peace footing Germany's armoured strength consisted of five armoured motorized divisions, four motorized divisions and four light divisions. An armoured division was made up of 345 heavy and medium tanks, and a light division was half that amount. It was these armoured machines that were going to lead the first lightning strikes into Poland.

Pz.Kpfw.II and Pz.Kpfw.I are seen here just prior to the invasion of Poland. Note the white crosses painted on the turrets for ground and aerial recognition. Each Panzer division had a tank brigade totalling some 324 tanks, of Pz.Kpfw.I through to Pz.Kpfw.IV types.

A long column of Pz.Kpfw.I. Although the Pz.Kpfw.I was under-armed and under-powered it was more than capable of combating Polish armoured vehicles.

A Horch cross-country vehicle passes a stationary Pz.Kpfw.I. Contrary to popular belief, during the first week of the invasion of Poland a number of German tank attacks were poorly coordinated with the accompanying infantry. It was not entirely easy for German commanders to put the new doctrine of Blitzkrieg into practice.

A Pz.Kpfw.II advances along a road bound for the front. Poland was found by German soldiers to be a land whose sprawling territory contained every type of terrain: hot, dry sandy areas, fertile plains as well as swamps, extensive forests, high mountain ranges and the main rivers that generally flowed north-south of the country and constituted a natural barrier against an east-west assault.

A Pz.Kpfw.35(t) advances through a Polish village during the furious armoured drive in Poland. Note the Panzer man is wearing his distinctive black Panzer uniforms and beret. The beret remained in service with the Panzer crews until January 1941.

Panzermen pause during the advance through Poland during the latter stages of the campaign. A P.zKpfw.I and a Pz.Kpfw.II can be seen parked in a field. Note the solid white cross on the Pz.Kpfw.II, which has been crudely removed by the crew. These white crosses were first ordered to be applied in August 1939. However, once these vehicles reached Poland, vehicle crews soon felt that the crosses were too prominent, and were providing a too easy aiming point for Polish anti-tank gunners even at longer than normal ranges.

A Pz.Kpfw.II can be seen on a dusty road somewhere in Poland. The Germans continued throughout the invasion to crush enemy defences, disrupting the logistic network and not being slow to use terror as an additional weapon. For the Poles, however, it was the beginning of the end. They were slowly withdrawing into a long, narrow pocket, within which they were eventually to be encircled, isolated and then destroyed.

A Panzer man perched in the turret of a Pz.Kpfw.I as it moves through a newly captured Polish town. In spite of these tanks being used extensively during the invasion it was soon obvious that the light tank did not have any combat potential. It suffered from an underpowered engine, uncompetitive armament, and was prone to heavy damage from anti-tank shells because of its thin armour plating.

A nice photograph showing Panzer men in the turret of their **Pz.Kpfw.I** advancing along a dusty road. Although these tanks were used in relatively large numbers against lightly armed opponents it suffered from very thin armour and had an inadequate main gun.

vo photographs during the victory
rade in Warsaw in early October
)39. Here Hitler takes the salute as
s armoured vehicles, comprising
z.Kpfw.II, move past the podium.
his victory parade symbolized the
ight of the German Army and its
)nquest over Poland in little more
an a month. They had won their
attle against Poland by
nplementing a series of
verwhelming, rapid penetrations.
hese penetrations were followed
/ the encirclement of an enemy
at, in contrast to its German
)unterparts, was bound by static
id inflexible defensive tactics.

Two Pz.Kpfw.II advancing across a field down a steep gradient on the Western Front in 1940. These vehicles belong to the 1st Panzer Division. Both in Poland and on the Western Front the new tried and tested Blitzkrieg strategy of warfare owed much to the German light tanks, in spite them being intended primarily for training and light reconnaissance work.

Two Pz.Kpfw.I advance through a destroyed town during the campaign on the Western Front in May or June 1940. During the Western campaign, in tank-versus-tank combat, the deficiency of the Pz.Kpfw.I soon became apparent. However, fortunately for the Wehrmacht, the light Panzers succeeded in destroying the bulk of the enemy armour using Blitzkrieg techniques.

A column of support vehicles and tanks, including a motorcyclist, advance through a captured French town. Divisional transport in a typical Panzer division in 1940 amounted to some 452 motorcycles, and 452 light and 1,133 heavy lorries. Each division relied heavily on wheeled transport in order to supply the armoured spearheads quickly and effectively.

A stationary Pz.Kpfw.I can be seen with its commander and Wehrmacht troops. An intrigued child from a local village clambers onboard the vehicle to take a look at this new revolutionary light tank.

Spread out in a field are stationary Horch cross-country vehicles and a halftrack. In the distance smoke rises in the air suggesting that the area has probably been attacked either by ground or aerial bombardment, before the column moved forward again.

A command vehicle moves forward, probably through the wooded area of the Ardennes in May 1940. For the Western Front campaign, the main German strike would be given to Army Group A, which would drive its armoured units through the Ardennes, and then swing round across the plains of northern France and make straight for the Channel coast, thereby cutting the Allied force in half and breaking the main enemy concentration in Belgium between Army Group A advancing from the south and Army Group B in the north. The task of Army Group B was to occupy Holland with motorized forces and to prevent the linking up of the Dutch Army with the Anglo-Belgian force.

On the Western Front, vehicles including troops move a PaK35/36 anti-tank gun advance through a field. The PaK35/36 became the standard anti-tank gun of the German Army during the early part of the war. It weighed only 432kg and had a sloping splinter shield. The gun fired a solid shot round at a muzzle velocity of 762m/s (2,500ft/s) to a maximum range of 4,025m.

Sitting with his commanders with maps in a field, Erwin Rommel can be seen. Behind him is a stationary Pz.Kpfw.38(t). His panzer force soon earned the name of the *Gespensterdivision* (Ghost/Phantom Division) because of its speed and the fact that not even the German High Command knew exactly where it was on the situation maps. Rommel had a 'lead from the front' attitude and would sometimes purposely cut communications with the High Command if he wished not to be disturbed. The 7th Panzer Division was one of the most successful divisions in the German arsenal in the Western Front campaign and covered vast distances in a short period of time. Hitler saw such tactics as showing complete confidence in the Blitzkrieg concept. However, Rommel was criticized by staff for being difficult to contact and locate.

Spread across a field on the march are various vehicles carrying troops towards the front during the Western Front campaign.

A PaK35/36 anti-tank crew preparing their gun for action, while a Pz.Kpfw.IV passes by at speed during some intensive action against an enemy contact.

Two copies of German post cards depicting fighting on the Western Front in 1940. Advancing in columns are various supporting vehicles and light and heavy tanks.

Pz.Kpfw.38(t) advance through a field supporting Wehrmacht troops as they go into action against an enemy contact. This light Czech-built tank became the most widely used and important light tank incorporated by the Panzertruppe during the early years of the war. For the campaign against the West there were some 264 of these machines distributed between some of the most powerful panzer divisions.

A photograph taken from the port hole of a Pz.Kpfw.38(t) showing another Pz.Kpfw.38(t) advancing into action. The vehicle was armed with a 3.7cm cannon, known in German service as the KwK 37(t). It was a semi-automatic falling block weapon that fired AP shot muzzle velocity of 750 metres per second and could easily penetrate 3.2cm of French or British armour at 1,100 metres.

A variety of light and heavy tanks advance across a field bound for the front.

A Pz.Kpfw.38(t), which has been dug in beneath a tree, breaks cover during the invasion of France in 1940. Of the 2,702 tanks fielded against the British and French, there were 264 Pz.Kpfw.38(t) distributed among the ten Panzer divisions.

A group of tanks comprising Pz.Kpfw.III and II are seen stationary in a field.

Pz.Kpfw.I crew pose for the camera with their machine stationary on a road. Although the armoured spearheads on the Western Front had been a complete success, it was in France that the Pz.Kpfw.I was totally outmoded as a battle tank.

The crew of a **Pz.Kpfw.II** are seen stationary on a road with their vehicle. A motorcycle combination passes them at speed. The primary task of the **Pz.Kpfw.II** was intended to be support of infantry and to fight other tanks. But although it was a well-built tank, in terms of armour, armament and mobility, it was not outstanding. However, on the Western Front in 1940 it proved its worth and was a highly successful light tank.

A **Pz.Kpfw.II** returning to its home station somewhere in Germany. Both the **Pz.Kpfw.I** and the **Pz.Kpfw.II** represented the substantial majority of the Panzers fighting in the West in 1940.

A well-camouflaged Pz.Kpfw.I Ausf.B on the Western Front. A member of the *Gebirgsjäger* or mountain soliders can be seen standing on the right with a pair of 6x30 binoculars.

In the foreground is a stationary Pz.Kpfw.II whilst in the background are Pz.Kpfw.IV and Pz.Kpfw.II, motorcycles and support vehicles. For the battle against the West there were some 366 Pz.Kpfw.IV that saw operations. Originally, the panzer was designed as an infantry support weapon, but by the end of the campaign in the West the tank had proved to be so diverse and effective that it earned a unique tactical role on the battlefield.

A German soldier armed with an MG34 on a bipod has positioned himself in a field while a Pz.Kpfw.II moves past in support of the march.

Inside a captured French town in June 1940 are a collection of various tanks comprising the Pz.Kpfw.I, II, III and IV. The Panzerwaffe was undoubtedly the backbone of the Blitzkrieg in the West, and to support its furious drive through the Low Countries and into the French heartlands the Panzers were supported by the infantry divisions. In northern France the front lines were shrinking, cracking slowly but surely under the massive German pressure. German units seemed to be progressing with an increased determination and vigour, convinced of their ability to crush the enemy before it could prepare a secondary line of defence.

A tank commander preparing to climb into his Pz.Kpfw.I. Throughout the battle of France the strategic onus of ground movement lay totally with German armour, and troops relied almost invariably upon it. The light panzers were proving successful across the entire front, in spite of losses.

A column of light tanks stationary beside a road. In total there were 2,072 tanks that invaded France and the Low Countries in 1940. There were 640 Pz.Kpfw.I, 825 Pz.Kpfw.II, 456 Pz.Kpfw.III, 366 Pz.Kpfw.IV, 151 Pz.Kpfw.35(t) and 264 Pz.Kpfw.38(t). The reserves comprised some 160 vehicles to replace combat losses and 135 Pz.Kpfw.I and Pz.Kpfw.II, which had been converted into armoured command tanks, which resulted in them losing their armament. The vehicles that had been distributed among the ten Panzer divisions were not distributed according to the formation of the battles they were supposed to perform.

A Pz.Kpfw.38(t) advances along a road. For the campaign in the West the 1 Panzer Division, 2 Panzer Division and 10 Panzer Division each comprised thirty Pz.Kpfw.I, one hundred Pz.Kpfw.II, ninety Pz.Kpfw.III and fifty-six Pz.Kpfw.IV. The 6 Panzer Division, 7 Panzer Division and 8 Panzer Division consisted of ten Pz.Kpfw.I, 132 Pz.Kpfw.35(t) or Pz.Kpfw.38(t) and thirty-six Pz.Kpfw.IV. A further nineteen Pz.Kpfw.35(t) were added to the 6 Panzer Division due to the compliment of a battery of sIG mechanized infantry guns. The 3 Panzer Division, 4 Panzer Division and 5 Panzer Division each consisted of 140 Pz.Kpfw.I, 110 Pz.Kpfw.II, fifty Pz.Kpfw.III and twenty-four Pz.Kpfw.IV.

Two Pz.Kpfw.II roll along the road bound for the front.

An interesting photograph showing **British POWs** being transported to the rear on board the engine deck of a Pz.Kpfw.I.

The crew of a Pz.Kpfw.38(t) have festooned logs on the engine deck of the vehicle due to some of the terrain it had to contend with during the Panzerwaffe's furious advance.

A crew member prepares to board his Pz.Kpfw.I inside a French village in June 1940. On the front lines in a number of areas German tank commanders reported that the enemy was simply brushed aside, thrown into complete confusion. In most cases the defenders lacked any force capable of mounting a strong coordinated counter-attack. British artillery eager to stem the tide of the German onslaught, poured a storm of fire into advancing German columns, but they soon found that the Germans were too strong to be brought to a halt for any appreciable length of time.

Two photographs showing the same decimated French town. Here light Panzers push forward. The quality of the German weapons, above all the tanks, was of immense importance to the Blitzkrieg. But also their tactics were the best: stubborn defence, concentrated local firepower from machine guns and mortars, and rapid counter-attacks by the panzers to recover lost ground. The invasion of the Low Countries and France was a product of great organization and staff work, and marvellous technical ingenuity.

Two Pz.Kpfw.II advance along a congested road during the campaign in the West. Note the national flags draped over the engine deck for aerial recognition.

A Pz.Kpfw.I Ausf.A has halted on a road after experiencing difficulties with the vehicle's drive wheel. A civilian has come to the aid of the crewman with water. The tracks of the Ausf.A had a ground pressure of only 780 g/sq.cm and gave the vehicle sufficient off-road capability.

Pz.Kpfw.II on a training ground. Much was owed to training panzer crews for the ensuing campaigns ahead. In France, out of the 825 Pz.Kpfw.II that fought, only 10 per cent were lost.

A Pz.Kpfw.II during a training exercise demonstrates its power and versatility by driving through a building. In order to protect its armament the 2cm cannon and MG34 machine gun have been removed.

A Pz.Kpfw.I on a muddy road. Although under-powered, with uncompetitive armament, and too thin an armour plate, 640 of these vehicles performed well against both French and British tanks.

A Wehrmacht soldier poses for the camera with a Pz.Kpfw.I, which is at a training ground in the summer of 1940.

A Pz.Kpfw.38(t) rolls along a dusty road in the summer of 1940. This light Czech-built tank became the most widely used and important light tank incorporated by the Panzertruppe during the early years of the war.

During the invasion of France a group of Pz.Kpfw.I and II from the I Panzer Division arrive at the Channel coast. The I Panzer Division consisted of two Panzer regiments, I and 2. Each regiment contained one hundred Pz.Kpfw.II. This was the largest contingent of any one type of Panzer in the entire division.

A Pz.Kpfw.I tank destroyer armed with a 4.7cm PaK(t) advances along a road. This armoured vehicle is attached to the Panzerjäger-Abteilung 570 and can be seen moving parallel to the railway line near the canal between Pommeroeul and Hensies in France.

A column of Pz.Kpfw.38(t) advance along a road. The main armament of this vehicle was the improved Skoda A7 3.7cm cannon, known in German service as the KwK 37(t).

A column of Pz.Kpfw.38(t) advance through a French town in May 1940. The crew is wearing the Zeltbahn or waterproof shelter triangle capes issued to the armed forces. This form of waterproof protective clothing was not used widely by Panzer crews, for the cape was difficult to manoeuvre inside the small confines of the tank.

In a French village, Pz.Kpfw.38(t) has fallen foul to enemy tank gunners, which have brought the Panzer to a flaming halt. An anti-tank shell has penetrated the vehicle's side with such considerable force it has immobilized it by blowing apart its track links. Black scorch marks over the wheels and hull suggest the vehicle may have had an internal fire.

A Pz.Kpfw.1 Ausf.A is stationary in a field in 1940, more than likely on a training exercise.

Pz.Kpfw.I and II can be seen in a familiar Panzer column stationary in a field. These vehicles still retain their original old dark grey colour. However, by this period of the war the old white crosses painted on the tanks had been removed and replaced with the new Balkenkreuz.

A column of Pz.Kpfw.I advance along a mud track during the campaign in France in May or June 1940.

A column of Pz.Kpfw.II pass through a village in 1940. The German attack through France was swift and effective and the French forces were shocked by the speed and fire power of their enemy.

A Pz.Kpfw.II stationary in Paris following the fall of the capital in June 1940. It was during the early evening of 14 June 1940 that the first German troops, men of the 9 Infantry-Division, entered Paris.

A good view of a Pz.Kpfw.I during a training exercise in 1940. At these training garrisons, there were large firing ranges and vast open areas where these vehicles could be used.

Chapter III
Barbarossa

For the German invasion of Russia, code-named *Barbarossa*, the German Army assembled some three million men, divided into a total of 105 infantry divisions and thirty-two Panzer divisions. There were 3,332 tanks, over 7,000 artillery pieces, 60,000 motor vehicles and 625,000 horses. This massive force was distributed into three German Army Groups. Army Group North, commanded by General Wilhelm Ritter von Leeb, had assembled its forces in East Prussia on the Lithuanian frontier. Leeb's force provided the main spearhead for the advance on Leningrad.

Army Group Centre, commanded by General Fedor von Bock, assembled on the 1939 Polish/Russian Frontier, both north and south of Warsaw. Bock's force consisted of forty-two infantry divisions of the 4th and 9th Armies and Panzer Groups II and III. Bock's army contained the largest number of infantry and Panzer divisions of the three army groups.

Army Group South, commanded by General Gerd von Rundstedt, was deployed down the longest stretch of border with Russia. The front, reaching from central Poland to the Black Sea, was held by one Panzer Group, three German and two Rumanian armies, plus a Hungarian motorized corps, under German command.

The Panzer divisions' main armour consisted of: 410 Pz.Kpfw.I, 746 Pz.Kpfw.II, 149 Pz.Kpfw.35(t), 623 Pz.Kpfw.38(t), 965 Pz.Kpfw.III, and 439 Pz.Kpfw.IV. This armoured force had to rely on obsolete light tanks to provide the armoured punch.

For the Russian offensive the Panzer divisions had been slightly modified in armoured firepower. They had been in fact diluted in strength in order to form the deployment of more divisions. The planners thought that by concentrating a number of Panzer divisions together they would be able to achieve a massive local superiority.

These new Panzer divisions contained one tank regiment of two, sometimes three *abteilungen* totaling some 150-200 tanks; two motorized rifle (*schützen*) regiments, each of two battalions, whose infantry were carried in armoured halftracks or similar vehicles; and a reconnaissance battalion of three companies (one motorcycle and two armoured cars). The motorized infantry divisions accompanying the Panzer divisions in the *Panzergruppe* were similarly organized, but were dangerously lacking in armoured support. The divisional artillery consisted of two field, one medium and one anti-tank regiment, and an anti-aircraft battalion. These were all motorized and more than capable of keeping up with the fast moving pace of the Panzers.

During the early morning of 22 June 1941, the German Army finally unleashed the maelstrom that was Barbarossa. After a month of victorious progress, the German armies were fighting on a front 1,000 miles wide. The Panzer divisions had exploited the terrain and concerted such a series of hammer blows to the Red Army that it was only a matter of time before the campaign would be over. Yet in spite of these successes the Panzer divisions were thinly spread out. Although the armoured spearheads were still achieving rapid victories on all fronts, supporting units were not keeping pace with them. Consequently, it became increasingly difficult to keep the Panzers supplied with fuel. And without fuel the drive would grind to a halt. Nevertheless, between June and late September 1941, the Panzer and motorized divisions were more or less unhindered by lack of supply, difficult terrain or bad weather conditions. However, on 6 October the first snowfall of the approaching winter was reported. It melted quickly, but turned the dirt roads into quagmires and rivers into raging torrents. The Russian autumn with its heavy rain, sleet and snow, had arrived. The Panzer divisions began to slow. Wheeled vehicles soon became stuck in a sea of mud and could only advance with the aid of tracked vehicles towing them. No preparations had been made for the winter and the Panzer divisions lacked the most basic supplies for cold weather. There were no chains available for towing vehicles, and no anti-freeze for the engines' coolant systems. Tank and infantrymen alike had no winter clothing either.

In blizzards and temperatures, which fell to thirty degrees below zero, the exhausted Panzer divisions soon ran out of fuel and ammunition, and were compelled to break off their attack within sight of Moscow. On 6 December all plans to capture the Russian capital in 1941 had to be abandoned.

By 22 December only 405 tanks were operational in front of Moscow with 780 out of action (but repairable). By the end of the year, the Germans reported a loss of 2,735 tanks plus 847 replacements since 22 June. The Panzer divisions were now reduced to less than 1,400 operational tanks.

By the end of 1941 the battle-weary divisions of the *Panzerwaffe*, which had taken part in Operation Barbarossa - were no longer fit to fight. Mobile operations had consequently ground to a halt. Fortunately for the exhausted Panzer crews and supporting units no mobile operations had been planned during the winter of 1941, let alone for 1942. In the freezing arctic temperatures the majority of the Panzer divisions were pulled out of their stagnant defensive positions and transferred to France, to rest, reorganize and retrain.

Alongside a river bank in April 1941 in the Balkans are Pz.Kpfw.II and a stationary halftrack.

In the Balkans in April 1941 are two Pz.Kpfw.III and a Pz.Kpfw.II, which have attempted to cross the fast-flowing river, but their weight has caused their tracks to sink into the bottom of the riverbed. Several Panzers that saw action in the Balkans had in fact been specially waterproofed for the projected invasion of the British Isles in 1940, although these tanks were obviously not.

In the Balkans the Gebirgsjäger, or Mountain Troops, have hitched a lift on board a Pz.Kpfw.II. This was often the quickest means of transporting foot soldiers from one part of the battlefront to another.

Two Pz.Kpfw.II, either Ausf.A or Ausf.B variants, belonging to the 1st Panzer Division, negotiate a steep gradient during the opening phase of the German invasion of the Soviet Union. During the summer of 1941 the Pz.Kpfw.II demonstrated that this light tank was so seriously under-gunned and under-armoured that it could not fight effectively on the Eastern Front.

An interesting photograph showing an Sd.Kfz.7 halftrack towing a disabled Pz.Kpfw.II Ausf.F across an improvised bridge. Both vehicles carry the markings belonging to Panzer Regiment 24 of the 24th Panzer Division. Note in the background stuck in water a Pz.Kfz.8 halftrack.

Advancing along a dirt track is Ausf.B Pz.Kpfw.II. Note the divisional insignia, an upright 'Y' with one vertical stroke, in yellow, on the superstructure front plate next to the driver's vision port-flap. This was used by the 8th Panzer Division during the invasion of Russia.

Panzers push forward along a road bound for the front during the opening stages of the invasion of Russia. The strongest army group in the German arsenal was Army Group Centre. This army group made a series of heavy penetrating drives through the Russian heartlands, bulldozing through the marshy ground to the main Russian defences. Within days of its first attacks across the frontier both the infantry and Panzer divisions had pulverized bewildered Russian formations, which led to a string of victories along its entire front.

In Russia a Pz.Kpfw.II advances along a typical Soviet road during the summer of 1941. The situation for the Russian defenders looked grim. The ferocity of the German attack was immense and without respite. Within a week of the invasion, almost continuous battle had left the Russian soldiers exhausted. Stalin's insistence that his troops must fight from fixed positions without any tactical retreat had caused many units to become encircled, leaving German tank units to speed past unhindered and achieve even deeper penetrations.

The crew of a Pz.Kpfw.38(t) pose for the camera onboard a railroad flat car for its trip to the Eastern Front in the summer of 1941. Note the chalk number on the bow plate, which is obscured by the crewman. This is probably the chassis number, which was often applied at the factory prior to issue to the receiving unit.

A column of Pz.Kpfw.II pass through a newly capture[d] Russian town. Watching the spectacle from the roadsi[de] are Jews. Unknown to them at the time, their days we[re] numbered. The Germans were to commit wholesa[le] murder, and transport many thousands to ghettos an[d] concentration camps across Europe and the Eas[t]

Trundling up a hill is a sIG 33 heavy infantry gun belonging to the 704 Gun company of the 5th Panzer Division. This vehicle's chassis has been converted on a Pz.Kpfw.I. The canvas sheeting attached to the front of the vehicle's shield was to protect the gun and the crew from dust, dirt and water. During the Russian campaign a number of sIG heavy infantry guns saw active service, but gradually most of the guns disappeared. Only the 5th Panzer Division retained some of these machines in active service until the summer of 1943.

A line of Pz.Kpfw.38(t) being transported during the early phase of the Russia campaign. They are being ferried across a river using just parts of a pontoon bridge. Provisions are being carried on the engine decks.

A column of Pz.Kpfw.I, probably in southern Russia in the summer of 1941. The tanks still retain their old familiar dark grey camouflage colour.

During a halt in the furious advance a Pz.Kpfw.I has halted in a field. One of the crew members can be seen looking at the engine.

Using a chain hoist attached to a log tripod, these engineers appear to be re-attaching the engine deck to a Pz.Kpfw.I Ausf.A. A standard feature on many Panzers was the removable engine-deck covering.

A column of Pz.Kpfw.38(t) moving towards the battlefront. The German attack along many parts of the Russian front was furious. As a consequence, Russian front units were simply brushed aside and totally destroyed by the Panzers. Red Army survivors recalled that they had been caught off guard, lulled into a false sense of security. Now they were being attacked by highly mobile armour and blasted by heavy artillery. In many places the force of attack was so heavy that they were unable to organize any type of defence. In total confusion, hundreds of troops, disheartened and frightened, retreated to avoid the slaughter, whilst other more fanatical units remained ruthlessly defending their positions to the death.

Three Panzers advance across rough terrain passing one of the many small villages that spread across the vast hinterland of the Soviet Union. The tank leading the advance is a Pz.Kpfw.38(t). Following closely behind is a Pz.Kpfw.II, and behind that is a Pz.Kpfw.IV.

An interesting photograph showing a dusty stationary Pz.Kpfw.II on a road somewhere in Russia. The sudden speed and depth of the German attack was a brilliant display of all-arms coordination. The Soviets were quite unprepared for the might of the German attack. In some areas along the front units were simply brushed aside by the Panzer divisions and totally destroyed.

A Pz.Kpfw.38(t) has been blown on its side by anti-tank shelling. The hole next to the Panzer is evidence of an explosion. A halftrack has arrived, probably to help salvage the machine.

A Pz.Kpfw.35(t) moves across a field towards the front line. Although this vehicle was under-gunned, during the early phase of the attack against Russia they proved relatively successful. As for the Red Army, in many places its force was unable to organize any sort of defence. In total confusion, thousands of troops, disheartened and frightened, retreated to avoid the slaughter, whilst other more fanatical units remained ruthlessly defending their positions to the death.

A Pz.Kpfw.II gives a helping hand with tow cable to help a stricken wheeled vehicle out of the mire during the early autumn of 1941. By this period of the year many roads had become boggy swamps. Although tracked vehicles managed to push forward through the mire at a slow pace, trucks and other wheeled vehicles were hopelessly stuck up to their axles in deep boggy mud. Within months the Soviet operation had gone from a rapid drive to a slow crawl.

A photograph of Pz.Kpfw.I stuck mud. The mud produced from a few hours of rain was enough to turn a typical Russian road into a quagmire. Often halftracks were called to help tow out stricken Panzers that had become mired. The halftrack was undoubtedly the workhorse on the Eastern Front, especially along the terrible road system that plagued seemingly endless miles of terrain. Its effective towing capability ensured that troops and ordnance often got through when they otherwise might not.

Two Pz.Kpfw.38(t) on a muddy road in Russia. The Soviet Union proved to be a completely alien environment to the Panzerwaffe, and the distances travelled soon proved to be more of a problem than ever imagined. Russia would not only test the endurance of the Panzer crews' physical stamina, but also their machines and supplies.

Locals, together with German soldiers, and crew are seen with a stationary Pz.Kpfw.38(t) in the winter of 1941.

A Pz.Kpfw.I supports an MG34 machine-gun crew advancing along a mountain road.

A Pz.Kpfw.II belonging to the 10th Panzer Division enters a newly captured Russian town in the early winter of 1941.

A Pz.Kpfw.II advances along rough terrain in the winter of 1941.

A column of vehicles, including Pz.Kpfw.38(t), on a snowy road bound for the front. The extreme winter of late 1941 had caused the German Army serious delays. As a result much of the front stagnated until the spring thaw of 1942, delaying the conquest of Russia by months. Barbarossa had been a success in terms of the vast distances in which the Wehrmacht had travelled but, coupled with growing enemy resistance and the Russian winter, it had failed to achieve its objective.

On a training ground in the snow, and a newly whitewashed Pz.Kpfw.I can be seen stationary next to a standard grey camouflage Pz.Kpfw.II. Note how the whitewashed vehicle is concealed against the snow terrain. At first virtually none of the German vehicles fighting on the Eastern Front received any type of winter whitewash paint and retained their original dark grey camouflage scheme, making them more susceptible to attack in the snow.

A **Pz.Kpfw.II** can be seen in a Soviet village in the winter of 1941. The vehicle has received a crude application of winter whitewash.

A Pz.Kpfw.38(t) that has received an application of winter whitewash is seen moving along a road. Horse carts can be seen passing. In Russia the bulk of all motive power was undertaken by animal draughts, and thousands of caissons and other pieces of equipment were required to support the drive east.

The crew of a Pz.Kpfw.I pose for the camera in the winter of 1941. The men wear their standard army issue greatcoat with woollen toque, which was a popular winter item during this period of the war. Scarves were sometimes worn with the toque for additional insulation.

In a Russian town are a line of parked Panzers comprising Pz.Kpfw.I, II, III and IV.

Total destruction on a road in Russia. These Pz.Kpfw.38(t) have more than likely been attacked by Soviet aircraft. On the road to Moscow alone, the Panzerwaffe reported the complete loss of 86 per cent of the Pz.Kpfw.38(t) force. These losses were so high that this model of Czech tank was never able to recover, and consequently they were phased out the following year.

A Pz.Kpfw.38(t) rolling along a road probably in the early summer of 1942 in southern Russia. In spite of the terrible problems that faced the badly depleted Panzer divisions, back in Germany in 1942 production of tanks continued to increase. In order to overcome the mammoth task of defeating the Red Army more Panzer divisions were being raised, and motorized divisions were converted into panzergrenadier divisions. Although equipping the Panzerwaffe was a slow expensive process, it was undertaken effectively with the introduction of a number of new fresh divisions being deployed on the front lines.

A Pz.Kpfw.38(t) moving across country. By the beginning of the German summer offensive in May 1942, not all the Panzer divisions on the Eastern Front were fully equipped and ready for combat. Some of the older units, for instance, did not have their losses from the winter offensive of 1941 replaced and were not ready for any type of full-scale operation. Worn out and depleted Panzer divisions were therefore relegated to Army Group North or Army Group Centre where they were hastily deployed for a series of defensive actions instead. The best equipped Panzer divisions were shifted south to Army Group South for operations through the Caucasus. It was entrusted to the 1st and 4th Panzer armies to spearhead the drive. By May 1942 most of the Panzer divisions involved were up to about 85 per cent of their original fighting strength, and had been equipped with Pz.Kpfw.III and Pz.Kpfw.IV.

Early evening and a Pz.Kpfw.38(t) advances along a sandy road in southern Russia in May 1942.

Two photographs showing a column of Pz.Kpfw.38(t) tanks advancing across Russia. During the early phase of the Russian campaign the Panzer divisions relied heavily on the lighter tanks, such as the Pz.Kpfw.38(t), to provide the armoured punch necessary to break through enemy lines. This gradually put an increasing strain on the light tanks, and as a result many of them were either destroyed or developed mechanical problems.

A Pz.Kpfw.II advances along a typical Russian dirt road. The distances which the Panzerwaffe had to travel were immense and supply lines were constantly being overstretched, especially by advanced units spearheading far in front of the main column.

A Pz.Kpfw.38(t) rolls along a road. Passing in the opposite direction is a column of Soviet prisoners being led to the rear to hastily erected POW camps. This Czech-built tank was relatively successful during early operations in Russia, however it was already clear that it was no match for the mighty Soviet T-34.

Two photographs showing the devastation caused to a German armoured column by Soviet aerial attack. Both photographs show damaged Pz.Kpfw.38(t). Often these vehicles were salvaged and their parts cannibalized for other tanks.

Early summer of 1942 and the crew of a Pz.Kpfw.38(t) can be seen sitting on the tank. It was quite obvious by this period of the war how lightly armoured these Czech vehicles were when it came to meeting the developing threat of the Soviet army. As a consequence losses grew to staggering proportions. By the time the Germans unleashed their summer offensive in May 1942, this light tank was being relegated to second line duties.

A Pz.Kpfw.38(t) wades across a river in 1942. In 1942 each Panzer Division fielded two motorized infantry regiments with two infantry battalions each.

On a congested road and a column of stationary Pz.Kpfw.35(t) can be seen, along with Horch cross-country vehicles. Note the national flag draped over the engine deck of the tank for aerial recognition purposes.

Climbing through hilly ground is an unidentified Panzer unit comprising Pz.Kpfw.I and II. The tank crews are wearing their special black Panzer uniforms of the *Panzertruppen*. Across Europe and into Russia these black uniforms would come to symbolize the elite troops armoured troops of the powerful Panzerwaffe.

A column of Pz.Kpfw.38(t) roll along a road. By mid-1942 many of the light tanks were being pulled from the front line and used as training vehicles. The losses in German armour during this period were immense and as a consequence heavier tanks were required to deal with Soviet tanks and guns.

A propaganda photograph showing the versatility of the Pz.Kpfw.38(t). While these vehicles were successful during the early period of the war and on the Eastern Front during the summer of 1941, they never had the status of a main battle tank.

An impressive photograph showing a very long column of Pz.Kpfw.38(t) rolling along the road in southern Russia, watched by bewildered Ukrainian women. Initially, the population of the Ukraine welcomed the German invasion. They saw the Germans as liberators from Soviet oppression.

Inside a square and a German armoured unit has halted. There are various vehicles, including Pz.Kpfw.II and III. A German commander can be seen watching the spectacle.

A crew member of a 15cm sIG33 can be seen relaxing in a seat on the side of a road, eating and drinking. This vehicle has been converted from the chassis of a Pz.Kpfw.I Ausf.B. Note the gun battery designation letter 'F' painted either in white or yellow on the side of the superstructure.

Chapter IV
Last years on the Eastern Front 1942-1943

In spite of the terrible problems that faced the badly depleted Panzer divisions, back in Germany production of tanks still increased. In order to overcome the mammoth task of defeating the Red Army more Panzer divisions were being raised, and motorized divisions converted into *Panzergrenadier* divisions. Although equipping the *Panzerwaffe* was a slow, expensive process, it was undertaken effectively with the introduction of a number of new fresh divisions being deployed on the front lines.

Another problem the *Panzerwaffe* were facing on the Eastern Front was heavier Russian tank fire power, for example their T35 tank. Hitler's light panzers were often no match against these vehicles and as a consequence a number of tanks were modified and up-gunned to deal with the developing threat. The Pz.Kpfw.I Ausf.B chassis mounted a heavier 15cm sIG33 gun, named the Bison. This mammoth piece was relatively short and barely fitted inside a tall superstructure. Although the tank had a powerful gun, it was not very versatile on the battlefield.

The Pz.Kpfw.II also saw its chassis get converted into what was known as a Marder II. These vehicles mounted the 7.62cm PaK 36(r), and later the 7.5cm PaK 40 was mounted on the tank chassis of the Ausf.F resulting in a better overall fighting machine. The Marder II became a key component in the *Panzerwaffe* arsenal and served with the Germans on all fronts through to the end of the war.

Another vehicle that served extensively on the Eastern Front was *Fahrgestell Panzerkampfwagen II* or Wespe. This mounted the 15cm sIG 33 gun, and there was a version for a 10.5cm *leichte Feldhaubitze 18/2* field howitzer in a built-up superstructure. The Panzer II proved an efficient chassis for this weapon and it became the only widely produced self-propelled 10.5cm howitzer for Germany. Between February 1943 and June 1944, 676 left the production lines and it served with German forces on all major fronts.

It was not just the Pz.Kpfw.I and II that were modified to meet the increasing enemy armoured threat, the Pz.Kpfw.38(t) and 35(t) were modified. The Pz.Kpfw.38(t) chasis was modified to mount the 7.5cm gun in an open-top superstructure known as the Sd.Kfz 138 or Marder III. Another variant, the Sd.Kfz 139 Marder III carried a Soviet 76.2cm gun in an open-top superstructure. Then there was the Sd.Kfz 138/1 Grille, which carried a German 150cm infantry gun. The Sd.Kfz 140 Flakpanzer 38(t) carried a 20mm anti-aircraft gun; and the SdKfz 140/1 *Aufklärungspanzer* 38(t) mit 2cm KwK 38 was a reconnaissance tank with a 2cm turret gun from a Sd.Kfz.222 armoured car, of which seventy were built. Then there was the Sd.Kfz 140/1 *Aufklärungspanzer* 38(t) mit 7.5cm KwK37 L/24, which mounted the 7.5cm gun in a modified superstructure, and later in the war the Jagdpanzer 38(t) left the production lines. This was unofficially known as the *Hetzer*, which was a powerful and deadly tank destroyer carrying a 7.5cm L/48 anti-tank gun.

The Pz.Kpfw.35(t) also saw its chassis being converted into a tank destroyer by substituting a captured Soviet 76.2cm field gun in its place. The gun and crew were protected by a thin, fixed, three-sided, partially roofed casemate that used armour plate salvaged from captured Soviet tanks. The prototype was completed by September 1943, although it used the older 76.2cm M-1936 F-22 field gun, and proved reasonably successful.

All these converted light panzers saw relative success on the battlefield. However, a number of them were still in development during 1942, and the factories were putting a lot of their time in rapidly building heavier panzers like the Pz.Kpfw.III and IV, and the new Tiger tank to meet the growing threat of heavier Russian armour.

By the beginning of the summer offensive in May 1942, not all of the Panzer divisions were fully equipped and ready for combat. Some of the older units, for instance, did not even have their losses from the winter offensive of 1941 replaced and were not ready for any type of full-scale operation. Worn out and depleted Panzer divisions were therefore relegated to Army Group North or Army Group Centre where they were hastily deployed for a series of defensive actions instead. The best-equipped Panzer divisions were shifted south to Army Group South for operations through the Caucasus. Two Panzer armies – 1st and 4th –were to spearhead the drive. By May 1942 most of the Panzer divisions involved were up to nearly 85 per cent of their original fighting strength, and had been equipped with Pz.Kpfw.III and Pz.Kpfw.IV.

With renewed confidence the summer offensive, codenamed 'Operation Blau', opened up in southern Russia. Some fifteen Panzer divisions and Panzergrenadier divisions of the 1st and 4th armies, together with Italian, Rumanian and Hungarian formations, crashed into action. In just two days the leading spearheads had pushed some 100 miles deep into the enemy lines and began to cut off the city of Voronezh. The city fell on 7 July. The two Panzer armies then converged with all their might on Stalingrad. It seemed that the Russians were now doomed. With an air of confidence Hitler decided to abandon the armoured advance on Stalingrad and embark on an encirclement operation down on the Don. The 6th Army was to go on and capture Stalingrad without any real Panzer support and fight a bloody battle of attrition there. Eventually, the fighting became so fierce it embroiled some twenty-one German divisions including six Panzer and Panzergrenadier divisions.

The 6th Army soon became encircled and three hurriedly reorganized under-strength Panzer divisions were thrown into a relief operation. By 19 December the 6th Panzer Division had fought its way to within 40 miles of Stalingrad. But under increasing Russian pressure the relief operation failed. The 6th Panzer Division and remnants of the 4th Panzer Army were forced to retreat, leaving the 6th Army in the encircled city to its fate. Some 94,000 soldiers surrendered on 2 February 1943. With them the 14th, 16th, and 24th Panzer Divisions, and the 3rd, 29th, and 60th Panzergrenadier Divisions were decimated.

The end now seemed destined to unfold, but still more resources were poured into the Panzer divisions. Throughout the early cold months of 1943, the *Panzerwaffe* built up the strength of the badly depleted Panzer divisions. These divisions consisted mainly of heavier tanks and tank destroyers like the Marder and Wesper. The Pz.Kpfw.I had ceased to be an effective battle tank and was delegated to training duties. Even the converted Bison could not achieve any success on the battlefield. The Pz.Kpfw.II and the Czech-built 38(t) and 35(t) were almost nonexistent. Only their converted variants now took prime position in the Panzer divisions.

By the summer of 1943, the *Panzerwaffe* fielded some twenty-four Panzer divisions on the Eastern Front alone. This was a staggering transformation of a Panzer force that had lost immeasurable amounts of armour in less than two years of combat. Hitler now intended to risk his precious *Panzerwaffe* in what became the largest tank battle of the Second World War, Operation *Zitadelle*.

Whilst Hitler's light panzers were never again to take pride of place on the battlefield, as they had done in the early years of war, their converted variants fought well for the next two years of combat. However, as with all the tanks and tank destroyers that fought in the once-vaunted *Panzerwaffe*, they were too few to make any significant difference on the Eastern Front, and as a consequence were destroyed.

A column of Pz.Kpfw.II Lynx tanks belonging to the 4th Panzer Division cross a snowy Russian landscape bound for the front lines. Note the stowage arrangement and modifications to the tank, which were typical of this unit during this period of the war.

Two Pz.Kpfw.II cross a pontoon bridge probably in December 1942 outside Stalingrad. In and around the decimated city of Stalingrad the fighting became so fierce it embroiled some twenty-one German divisions including six Panzer and Panzergrenadier divisions. The 6th Army soon became encircled and three hurriedly reorganized under-strength Panzer divisions were thrown into a relief operation. By 19 December the 6th Panzer Division had fought its way to within 50km of Stalingrad. But under increasing Russian pressure the relief operation failed. The 6th Panzer Division and remnants of the 4th Panzer Army were forced to retreat, leaving the 6th Army in the encircled city to its fate. Some 94,000 soldiers surrendered on 2 February 1943. With them the 14th, 16th, and 24th Panzer Divisions, and the 3rd, 29th, and 60th Panzergrenadier Divisions were decimated.

During the winter, a Pz.Kpfw.38(t) can be seen poised on an icy road somewhere on the Eastern Front. Note the dented locker on the tank. In spite of the reverses on the Eastern Front, throughout the early cold months of 1943, the Panzerwaffe drastically tried to build up the strength of the badly depleted Panzer divisions.

Pioneers watch as a Pz.Kpfw.II crosses a pontoon bridge in Russia. Foliage can be seen attached to the rear of the vehicle. By this period of the war the light tanks were already dwindling in quantity due to high losses sustained against heavier enemy armour like the T-34.

One of the crew members of a Pz.Kpfw.38(t) is about to dismount from his stationary tank. The vehicle appears to be heavily laden with supplies for the long duration of the march. This tank belongs to the 12th Panzer Division.

A crew member standing next to his Pz.Kpfw.38(t) Ausf.B, C or D variant, as identified by the bolt pattern on the turret front plates for 25mm armour. This vehicle belongs to the 12th Panzer Division. Note the stowage arrangement.

The crew of a Pz.Kpfw.38(t) Ausf. B, C or D gets some air as the tank advances along a road. This unit has applied a black/white national cross as well as a three-digit tactical number, '223', in yellow, to the near side locker.

A Pz.Kpfw.38(t) leads a column of the heavy Pz.Kpfw.IV bound for the front lines.

A pair of Pz.Kpfw.II advance through a newly captured village somewhere on the Eastern Front. They lead several Pz.Kpfw.38(t). These vehicles belong to the 12th Panzer Division.

On a congested road are **Pz.Kpfw.II** and **III** accompanied by **Sd.Kfz.251** halftracks.

Crossing a makeshift bridge is a column of **Marder III** armed with the captured 7.62cm Russian anti-tank gun. This vehicle had a lightly armoured three-sided shield fitted directly onto the **Pz.Kpfw.38(t)** chassis.

Two Pz.Kpfw.II out in the field in Northern Russia in 1942. Waffen-SS troops belonging to the 4th SS Polizei Division can be seen standing next to the leading vehicle.

A column of Pz.Kpfw.II advance through a captured Russian village bound for the front.

Passing a group of Russian peasants, a Wespe advances towards the front lines probably in the summer of 1943. The production of the Wespe tank began in 1942. It was armed with a 10.5cm light field howitzer in an open-topped, box-like superstructure built on top of a Pz.Kpfw.II chassis. Between 1942 and 1944, 683 of these self-propelled guns were built.

July 1943 and the crew of a Marder III Panzerjäger pose for the camera. The Marder III was the first of a series of improvised light tank hunters and was built on the chassis of a Pz.Kpfw.38(t). This particular vehicle is fitted with a captured 7.62cm Russian Model 36 anti-tank gun.

During the battle of Kursk in July 1943 a well-camouflaged Wespe battery fords a shallow stream. The Wespe was thinly armoured to offset the added weight of the gun and, like all self-propelled gun mounts, had open hulls that offered little protection for the men and guns.

British POWs pass two stationary vehicles. One is a Pz.Kpfw.II, and parked behind is a Sturmpanzer or Brummbar (Grizzly bear). This was a tank with an assault howitzer for use in urban combat, designed to be mounted on a Pz.Kpfw.IV chassis. Note the application of Zimmerit anti-magnetic mine paste.

A battery of Wespe self-propelled guns open fire simultaneously during defensive operations in Poland in the late summer of 1944. Both vehicle's guns can be seen recoiling. The sound of these guns in action can be easily imagined, and is evidenced by a number of the crew holding their ears.

A Marder III somewhere on the Eastern Front, probably in 1942. This tank destroyer was built on the chassis of the Pz.Kpfw.38(t). It was armed with the captured Soviet 76.2mm F-22 M 1936 divisional gun, or the German 7.5cm PaK 40.

A Marder II self-propelled anti-tank gun advances along a muddy field. Because the Germans had few anti-tank weapons cable of successfully engaging the Soviet T-34 and KV heavy tanks, an urgent need arose for a more mobile and powerful anti-tank weapon. The Marder II was built on the chassis of a Pz.Kpfw.II.

A Marder out on the battlefield accompanied by an Sd.Kfz.251 halftrack. The Marder series was not fully armoured. It had thin upper armour protection and all were open tops. Some were issued with canvas tops to protect the crews from the weather.

Taking cover in the undergrowth with a full complement of crew is a Marder II. This photograph was probably taken in Italy in 1943.

A photograph of a Wespe self-propelled light field howitzer in undergrowth, probably during the Kursk offensive in July 1943. Built on the modified chassis of the Pz.Kpfw.II, the Wespe mounted the 10.5cm leFH18/2 and was issued to the Panzer artillery regiments of the Panzer and Panzergrenadier divisions.

A well-camouflaged Marder hurtles along a road bound for the front line in the summer of 1943.

During the Kursk offensive and SS troops can be seen surveying a destroyed Soviet T-34 tank. In the background is a stationary Wespe, probably belonging to an SS unit.

A late new Jagdpanzer 38(t) manufactured at the Skoda plant. A total of 2,584 of these vehicles were constructed on the chassis of a Pz.Kpfw.38(t) from April 1944 to May 1945. The vehicle has received a camouflage paint scheme of red and brown and olive green patches over the dark yellow sand base.

APPENDIX 1
Panzer I variants

Panzerkampfwagen I Ausf.A ohne Aufbau

The first Panzer I vehicles to be built. Fifteen of this variant were completed by various firms: Daimler-Benz, Henschel, Krupp, MAN, and Rheinmetall. The Ausf.A ohne Aufbau was a Panzer I hull without any superstructure or turret. The interior was completely open. The suspension and hull were identical to the Ausf.A, but total weight was reduced to 3.5 tons and height to 1.15m. Performance was similar.

Munitionsschlepper auf Panzerkampfwagen I Ausf.A

Fifty-one Sd.Kfz 111, the *Munitionsschlepper* (ammunition tractor) were built to provide Panzer units with an armoured tracked vehicle in order to resupply panzer units on the front.

Brückenleger auf Panzerkampfwagen I Ausf.A

Bridging equipment was tested on prototype Ausf.A chassis. However, due to the weak suspension it was decided to use it on the Panzer II chassis.

Flammenwerfer auf Panzerkampfwagen I Ausf.A

Designated as the Sd.Kfz 265, the kl.Pz.BefWg *Flammenwerfer* had a portable flamethrower and was mounted in one of the machine guns. It was intended to give the Panzer I additional firepower against close targets.

Kleiner Panzerbefehlswagen (kl.Pz.Bef.Wg)

The kl.Pz.Bef.Wg served with all Panzer units during the early part of the war. One hundred and eighty-four were built by Daimler-Benz at the same time as Ausf.B production, and six prototypes were built from Ausf.A Panzers.

4.7cm PaK(t) (Sf) auf Panzerkampfwagen I Ausf.B

Known as the Panzerjäger I, or tank hunter, this was the first in a series of Panzer destroyers. It comprised of a mounted Czech 4.7cm PaK(t) antitank gun. Eighty-six rounds were carried for the main gun.

15cm sIG 33 (Sf) auf Panzerkampfwagen I Ausf.B

Known as the Bison it mounted a 15cm heavy SIG33 infantry gun. Thirty-eight of these vehicles were converted from Ausf.B variants in February 1940. They served with six heavy SP infantry gun companies, and remained in service until 1943.

Flammenwerfer auf Panzerkampfwagen I Ausf.B

This prototype field model was constructed, but there is no record of it entering service.

Ladungsleger auf Panzerkampfwagen I Ausf.B

The *Ladungsleger*, or explosives layer, was converted on the rear deck of an Ausf.B Panzer and used to lay explosives to destroy field fortifications. Whilst rare, these vehicles were used in armoured engineer companies.

Flakpanzer

The Panzer I was converted into a self-propelled anti-aircraft gun, and known as a Flakpanzer I. Due to high manufacturing costs it was rarely seen in service.

Panzer II Variants

Panzer II Ausf.A (Pz.Kpfw.IIA)

The Ausf.A was built in limited numbers, and was divided into three sub-variants. The Ausf.A/1 was initially built with a cast idler wheel with rubber tyre, but this was replaced after ten production prototypes with a welded part. The Ausf.A/2 had better engine access. The Ausf.A/3 included improved suspension and engine cooling. In total seventy-five were produced from May 1936 to February 1937 by Daimler-Benz and MAN.

Panzer II Ausf.B (Pz.Kpfw.IIB)

The Ausf.B saw limited service. It had new suspension, the length of the body was increased, and there was increased superstructure, deck, and turret roof strength. Only twenty-seven of these vehicles were produced.

Panzer II Ausf.C (Pz.Kpfw.IIC)

The Ausf.C was the last in the first series of Pz.Kpfw.II. A number of modifications were made including the suspension, with the replacement of the six small road wheels and tracks. Some twenty-five of these variants were produced from March until July 1937.

Panzer II Ausf.A, B and C

This was the first real production model. The Ausf.A entered production in July 1937 and was superseded by the Ausf.B in December 1937. A few minor changes were made in the Ausf.C version, which became the standard production model from June 1938 through to April 1940. A total of 1,113 of these vehicles were built from March 1937 to April 1940 by Alkett, FAMO, Daimler-Benz, Henschel, MAN, MIAG, and Wegmann.

Panzer II Ausf.D and E

Ausf.D was developed for a reconnaissance role. Only the turret was the same as the Ausf.C model, with a new hull and superstructure design and the use of a Maybach HL62TRM engine driving a seven-gear transmission (plus reverse). A total of 143 Ausf.D and Ausf.E tanks were built from May 1938 to August 1939 by MAN, and they served in Poland. They were withdrawn in March 1940 for conversion to other types after proving to have poor off-road performance.

Panzer II Ausf.F

The Ausf.C, the Ausf.F were designed as reconnaissance tanks and served in the same role as the earlier models. 524 were built from March 1941 to December 1942 as the final major tank version of the Panzer II series.

Panzer II (Flamm)

The Ausf.D and Ausf.E tank variant, known as the Flamm and nicknamed the Flamingo had a new turret and mounted a single MG34 machine gun. Two remotely controlled flamethrowers were mounted in small turrets at each front corner of the vehicle. 150 of these were built from January 1940 to March 1942.

Panzer II Ausf.L "Luchs"

This light reconnaissance tank was designated as Ausf.L. It was the only Panzer II design with the *Schachtellaufwerk* overlapping road wheels and "slack track" configuration to enter series production. Some 100 of these vehicles were built from September 1943 to January 1944 as well as the conversion of four Ausf.M tanks. It became known as the *Panzerspähwagen* II, and was nicknamed "Lynx".

Self-propelled guns on Panzer II chassis

The 15cm sIG 33 auf Fahrgestell Panzerkampfwagen II (Sf) mounted a 15cm sIG33 heavy gun on a on a turretless Panzer II chassis. It had an open-topped 15cm thick armoured superstructure to protect it against small arms and shrapnel. Only twelve were built in November and December 1941. These served with the 707th and

708th Heavy Infantry gun companies in North Africa until they were finally destroyed in action in 1943.

Another self-propelled gun was built on the chassis of the Panzer II. Mounted on its superstructure was the 7.62cm PaK 36(r) auf Fahrgestell Panzerkampfwagen II Ausf.D/E (Sd.Kfz. 132). The gun was a captured Soviet 76.2cm anti-tank gun.

7.5cm PaK 40 auf Fahrgestell Panzerkampfwagen II (Marder II) (Sd.Kfz. 131)

The 7.5cm PaK 40 was mounted on the chassis of the Ausf.F. 576 of these models were built from June 1942 to June 1943. The Marder II became a major tank destroyer on the Eastern Front and saw action until the end of the war.

5cm PaK 38 auf Fahrgestell Panzerkampfwagen II

A 5cm PaK 38 was mounted on the Panzer II chassis. It was only moderately successful and so not many of these variants were produced.

Leichte Feldhaubitze 18 auf Fahrgestell Panzerkampfwagen II (Wespe)

This vehicle mounted a 10.5cm *leichte Feldhaubitze 18/2* field howitzer in a built-up superstructure. The Panzer II proved an efficient chassis for this weapon and it became the only widely produced self-propelled 10.5cm howitzer in the German arsenal. Between February 1943 and June 1944. Six hundred and seventy-six were built, by FAMO. It served on all major fronts and was nicknamed the Wasp

Munitions Selbstfahrlafette auf Fahrgestell Panzerkampfwagen II

To support the Wespe on the battle front, a number of Wespe chassis were completed without mounting the howitzer, and were used as ammunition carriers instead. They carried ninety rounds of ammunition. One hundred and fifty-nine were produced, and could be converted by mounting the leFH 18 in the field if required.

Limited production, experiments and prototypes

Panzer II Ausf.G (Pz.Kpfw.IIG)

Known as the *Schachtellaufwerk* by the Germans, this was a reconnaissance tank. Only twelve were ever built and it is not known if they ever reached active service.

Panzer II Ausf.H (Pz.Kpfw.IIH)

Only prototypes were ever produced and these were cancelled in September 1942.

Brückenleger auf Panzerkampfwagen II

These were built for bridge laying and served with the 7th Panzer Division in May 1940. Another vehicle was produced, designated as the Panzer II Ausf.J (Pz.Kpfw.IIJ), which included heavier armour. Twenty-two were produced by MAN between April and December 1942, and seven were deployed to the 12th Panzer Division on the Eastern Front.

Bergepanzerwagen auf Panzerkampfwagen II Ausf.J

A Panzer II recovery vehicle, about which little is recorded.

Panzer II Ausf.M (Pz.Kpfw.IIM)

The Ausf.M replaced the turret with a larger, open-topped turret mounting a 5cm KwK 39/1 gun. Four were built by MAN in August 1942, but did not see service.

Panzerkampfwagen II ohne Aufbau

A number of chassis of these vehicles were used for engineers, for personnel and as equipment carriers.

Panzer Selbstfahrlafette 1C

This prototype Panzer II chassis mounted a 5cm PaK 38 gun, on the chassis of the Ausf.G. Only two of these vehicles were produced. Both saw active service.

VK 1602 Leopard

The VK 1602 was intended as a replacement for the Ausf.L, armed with a 5cm KwK 39, powered by a Maybach HL157P engine, driving an eight speed transmission (plus reverse). While the hull was based on that of the Pz.Kpfw.IIJ, it was redesigned after the Pz.Kpfw.V Panther. However, neither of them saw service.

Panzer 38(t) Variants

Panzer 38(t) Ausf.A-C

General:
  Role: Medium tank
  Crew: 4

Armament and armour:
  Main armament: 37.2mm Skoda A7 (L/47.8) gun with 90 rounds

Secondary armament:
  2 x 7.92mm MG 37(t) (Model 37) machine gun with 2,550 rounds.

Armour:
  front 25mm, side 15mm

Power and weight:
  Power/Weight: 10 kW/metric ton (13.0 hp/short ton)
  Power: 91.9 kW (123.3 hp, 125 PS)
  Engine: Praga EPA Model I inline six-cylinder, liquid-cooled, petrol
  Bore: 110mm (~ 4.331 in)
  Stroke: 136mm (~ 5.354 in)
  Displacement: 7754.7 cc (~ 473.22 cu in)

Transmission:
  5 forward, I reverse

Weight:
  combat: 9.5 tonnes, dry: 8.5 tonnes

Performance:
  Speed: 56km/h (35mph)
  Range: 200km (120miles)

Dimensions:
  Length: 4.61m

Width: 2.14m
Height: 2.40m

PzKpfw 38(t) Ausf.A-D

TNH tank in German manufacture.

PzKpfw 38(t) Ausf.E-G

Pz 38(t) with frontal armour increased to 50mm by bolting on an additional 25mm armour.

PzKpfw 38(t) Ausf.S

90 TNH ordered by Sweden in February 1940 but seized by Germany.

Stridsvagn m/41 S(eries)I

Swedish license-built TNH version as compensation for the seized Ausf.S tanks. One hundred and sixteen produced.

Stridsvagn m/41 S(eries)II

Strv m/41 with upgraded armour and stronger engine. One hundred and four produced.

SdKfz 138 Marder III

Carried German 75mm gun in open-top superstructure.

SdKfz 139 Marder III

Carried Soviet 76.2mm gun in open-top superstructure.

SdKfz 138/1 Grille

Carried German 150mm infantry gun; also munition variant, which carried ammunition.

SdKfz 140 Flakpanzer 38(t)

Carried a 20mm anti-aircraft gun.

SdKfz 140/1

Aufklärungspanzer 38(t) mit 2cm KwK 38 reconnaissance tank with 20mm turret from a SdKfz.222 armoured car (seventy built).

SdKfz 140/1

Aufklärungspanzer 38(t) mit 7.5cm KwK37 L/24, 75mm gun mounted in a modified superstructure (only two built).

Jagdpanzer 38(t)

(Unofficially known as the *Hetzer*) a tank destroyer carrying a 75mm L/48 anti-tank gun.

G-13

Swiss designation for post-war-built Jagdpanzer 38(t) sold by Czechoslovakia.

Nahkampfkanone 1

Swiss-built tank destroyer, similar to Marder III. Only one built.

Pansarbandvagn 301

Swedish Stridsvagn m/41 (SI and SII). In the 1950s, 220 were rebuilt to armoured personnel carriers.

Stormartillerivagn m/43

Assault gun based on the m/41 SII chassis. Thirty-six produced.

APPENDIX 2
Armoured crew uniforms

Wearing their special black Panzer uniforms the *Panzertruppen* were very distinctive from the German soldier wearing his field-grey service uniform. The uniform was first issued to crews in 1934, and was the same design and colouring for all ranks of the Panzer arm, except for some of the rank insignia and national emblems worn by officers and generals. The colour of the uniform was specially dyed in black to hide oil and other stains from the environment of working with the armoured vehicles. These black uniforms would symbolize a band of elite troops that gained notorie*ty across Europe and into Russia for spearheading the powerful Panzerwaffe.

The black Panzer uniform itself was made of high-quality black wool, which was smooth and free of imperfections. The uniform comprised a short black double-breasted jacket worn with loose-fitting black trousers. The deeply double-breasted jacket was high-waisted and was specially designed to allow the wearer to move around inside his often cramped vehicle with relative comfort. The trousers were designed to be loose also in order to enable the wearer plenty of movement.

The 1934 pattern Panzer jacket was only in production until it was replaced in 1936 by the second pattern. This pattern was popular and remained in production throughout the war. It was very similar to that of the first pattern. It had the short double-breasted jacket, which was normally worn open at the neck, showing the mouse-grey shirt and black tie, but it also had provision for buttoning and hooking the collar closed for protection against weather.

On the jacket the shoulder straps, collar patches and around the death head skull were piped in rose pink *Waffenfarbe* material. The rose pink piping was worn by all ranks around the outer edge of the jacket collar, but this design was discontinued by 1942. Members of the 24th Panzer Division did not wear the rose pink piping, their piping was gold yellow. This colour piping was purely for commemorative wear and had been originally worn by the 1st Kavallerie-Division, which was the only cavalry division in the German Army to be converted to a fully-fledged Panzer division.

The German national emblem on the double-breasted Panzer jacket was very similar to that worn on the German service uniform. It was stitched on the right breast in heavy white cotton weave, but the quality and colour varied according to rank. They were also manufactured in grey cotton yarn or in fine aluminum thread. For officers and generals of the Panzertruppen they were normally heavily embroidered in silver wire.

The jacket was specially designed so that the number of buttons worn on the outside of the coat was limited, although there were two small black buttons positioned one above the other on the right side of the chest. These were stitched into place to secure the left lapel when the jacket was closed up at the neck.

The trousers worn were identical for all ranks. There was no piping used on the outer seams of the trouser legs. Generals of the Panzertruppen did not wear the red stripe on the trousers, as they did with the German Army service uniform. The trousers had two side pockets with button-down pocket flaps, a fob pocket and a hip pocket. The trousers were generally gathered around the tops of the short leather lace-up ankle boots.

The headgear worn by the Panzer crews in 1941 was the Panzer enlisted man's field cap or *Feldmütze* and was worn by all ranks. It was black and had the early type national emblem stitched in white on the front on the cap above a woven cockade, which was displayed in the national colours. The field cap had a pink soutache. For the next three years of the war the Panzer arm extensively wore the Panzer field cap. However in 1943 a new form of head dress was introduced, the *Einheitsfeldmütze* – the Panzer enlisted man's model 1943 field cap. The M1943 cap was issued in black, but when stocks ran low troops were seen wearing field-grey field caps. Both colours of the design were worn universally among Panzer crews and the cap insignia only slightly differed between the various ranks.

The field-grey German Army steel helmet was also issued to the Panzertruppen as part of their regulation uniform. Generally, the steel helmet was not worn inside the cramped confines of a tank, except when crossing over rough terrain and normally when the crewmember was exposed under combat conditions outside his vehicle. Many crews, however, utilized their steel helmets as added armoured protection and attached them to the side of the tanks cupola, and to the rear of the vehicle.

Another item of headgear worn by the Panzer arm was the officer's service cap or *Schirmmütze*. Although this service cap was not technically an item designed for the Panzer arm, it was none the less an integral part of the Panzer officer's uniform and was worn throughout the war.

The Panzer uniform remained a well-liked and very popular item of clothing and did not alter extensively during the war. However, in 1942 a special two-piece reed-green denim suit was issued to Panzer crews in areas of operations where the climate was considered warmer than normal theatres of combat. The new denim suit was hard wearing, light and easy to wash, and many crews were seen wearing the uniform during the summer months. The uniform was generally worn by armoured crews, maintenance, and even Panzergrenadiers who were operating with half-tracked vehicles, notably the Sd.Kfz.251 series. This popular and practical garment was identical in cut to the special black Panzer uniform. It consisted of the normal insignia, including the national emblem, Panzer death head collar patches and shoulder straps.

Apart from the uniforms worn by the Panzer crews, a special uniform was introduced for both *Sturmartillerie* and *Panzerjäger* units. The uniform was specially designed to be worn primarily inside and away from the troops' armoured vehicles, and for this reason designers had produced a garment that gave better camouflage qualities than the standard black Panzer uniform. The uniform worn by units of the *Panzerjäger* was made entirely from lightweight grey-green wool material. The cut was very similar to that of the black Panzer uniform. However, it did differ in respect of insignia and the collar patches.

The *Panzerjäger* uniform was a very practical garment. The cut was identical to that of the *Sturmartillerie* uniform, but was of a different colour. The uniform was made entirely of field-grey cloth, but again differed in respect of certain insignia. The collar patches consisted of the death's head emblems, which were stitched on patches of dark blue-green cloth and were edged with bright red *Waffenfarbe* piping. Officers did not display the death's head collar patches, but wore the field service collar patches instead. Also there was no piping on the collar patches.

Like the summer two-piece reed-green denim suit worn by Panzer crews, both tank destroyer and self-propelled assault gun units also had their own working and summer uniforms, which were also produced in the same colour and material.

Apart from the basic issued items of clothing worn by crews of the Panzer, tank destroyer and self-propelled assault gun units, crews were also issued with various items of clothing to protect them against the harsh climates. By the winter of 1942-43 the German Army had developed a new revolutionary item of clothing for the armoured crews called the Parka. The Parka was a well-made item of clothing that was well-padded and kept crews warm. Initially, the Parka was designed in field-grey with a reversible winter white. But by late 1943 a new modification was made by replacing the field-grey side with a camouflage pattern, either green splinter or tan-and-water. The coat was double breasted with the interior set of buttons being fastened to provide additional protection.

APPENDIX 3
Camouflage

During the invasion of Poland, the Low Countries, France and the invasion of the Soviet Union in June 1941, virtually all German equipment was painted in dark grey. During the invasion there were literally thousands of vehicles distributed between the Panzer divisions.

For the first four months of operation Barbarossa the vehicles painted in their overall dark-grey camouflage scheme blended well against the local terrain. However, with the drastic onset of winter and the first snow showers at the end of October 1941, Panzer crews would soon be filled with anxiety, as their vehicles were not camouflaged for winter warfare. With the worrying prospects of fighting in Russia in the snow the Wehrmacht reluctantly issued washable, white winter camouflage paint in November 1941. The paint was specially designed to be thinned with water and applied to all vehicles and equipment when snow was on the ground. This new winter whitewash paint could easily be washed off by the crews in the spring, exposing the dark grey base colour underneath. Unfortunately for the crews the order came too late and distribution to the front lines was delayed by weeks. Consequently, the crews had to adapt and find various crude substitutes to camouflage their vehicles. Some hastily applied their vehicles with a rough coat of limewash, others used lumps of chalk, white cloth strips and sheets, or even hand-packed snow, in drastic attempts to conceal conspicuous dark grey parts. Other vehicles, however, roamed the white arctic wilderness with no camouflage at all.

Following the harsh winter of 1941, the spring of 1942 saw the return of the dark grey base colour on most vehicles, while others returned to pre-war dark brown and dark green camouflage schemes. Crews had learnt from the previous year the lessons of camouflage. Survival for these young men was paramount. Many crews began adding to their camouflage schemes by finding various substitutes and applying them to the surface of their vehicles, such as foliage and bundles of grass and hay. This was a particularly effective method as it would break up the distinctive shapes of vehicles and allow them to blend into the local terrain. Mud, too, was used as an effective form of camouflage but was never universally adopted by the crews.

For the first time in southern Russia, in the Crimea and the Caucasus, where the summer weather was similar to that in North Africa, many vehicles were given an application of tropical camouflage, with the widespread use of sand colour schemes, almost identical to those used in the *Afrika-Korps*, or the tropical colours of yellow-brown RAL 8000, grey-green RAL 7008, or plain brown RAL 8017.

By 1943, olive green was being used on vehicles, weapons, and large pieces of equipment. A red-brown colour RAL 8012 also had been introduced at the same time. These two colours, along with a new colour base of dark yellow RAL 7028, were issued to crews in the form of a concentrated paste. Pastes arrived in 2kg and 20kg cans, and units were ordered to apply them over the entire surfaces of their vehicles. The paste was specially adapted so that it could be thinned with water or even fuel, and could be applied by spray, brush or mop.

The dark yellow paste was issued primarily to cover unwanted colours or areas of camouflage, especially during changes in seasons. These new variations of colours gave the crews the widest possible choices in schemes so as to blend in as much as possible to the local terrain. The pastes were also used to colour canvas tops and tarpaulins on the vehicles.

The new three-colour paint scheme worked very well on the front lines and allowed each unit to camouflage in many different landscapes. However, within months there were frequent problems with supply. Support vehicles carrying the new paste had to travel so far to various scattered units, even from railheads, that frequently Panzer units never received any new application of camouflage. Another problem was that many Panzer units were already heavily embroiled in bitter fighting and had neither the vehicles to spare nor manpower to pull them out for a repaint. Even rear area ordnance workshops were returning vehicles to action at such speed that parts might be replaced, but repainting omitted. A great number of vehicles never received any paste colours at all, and those that fought on remained in dark yellow, sometimes with crews adapting and enhancing colour schemes with the application of foliage and mud.

However, of all the failings, the greatest of them all was the actual paints themselves. These proved to be unstable when mixed with water, and even the lightest rain could cause colours to run or wash off vehicles. Even fuel, which was used to give the paste a durable finish, was at such a premium during the later stages of the war, that units were compelled to use water, waste oil and mixed or other paints. All this caused wide variations in the appearance of the paint schemes and as a consequence there were unusual colours like brick red, chocolate brown and light green. In spite of these variations in colour, and the fact that there had become little standardization in the camouflage schemes, occasionally there were complete units that appeared on the front lines properly painted and marked. But this was a rare occurrence.